Anonymous

A Catechism of Geography

Being an easy introduction to the knowledge of the world and its

inhabitants, the whole of which may be committed to memory at an early

age. Eleventh Edition

Anonymous

A Catechism of Geography
Being an easy introduction to the knowledge of the world and its inhabitants, the whole of which may be committed to memory at an early age. Eleventh Edition

ISBN/EAN: 9783337256319

Printed in Europe, USA, Canada, Australia, Japan

Cover: Foto ©berggeist007 / pixelio.de

More available books at **www.hansebooks.com**

A CATECHISM

OF

GEOGRAPHY:

BEING AN

EASY INTRODUCTION

TO THE

Knowledge of the World and its Inhabitants,

*The whole of which may be committed to memory
at an early age.*

ELEVENTH EDITION.

MONTREAL:
CHARLES G. DAGG.
1860.

A CATECHISM OF GEOGRAPHY.

CHAPTER I.

Definitions.

Question. What is GEOGRAPHY ?
Answer. A description of the earth.
Q. What is the earth ?
A. The world, or rather the globe, on which we live.
Q. Of what shape is the earth ?
A. nearly round, like an orange. *
Q. How large is the earth ?
A. It is more than twenty-four thousand miles round, and eight thousand through. †
Q. How far is it from the sun ?
A. Ninety-five millions of miles.
Q. Does the earth move ?
A. Yes ; it has two motions ; the one round the sun, which it performs yearly ; and the other round its own axis, which it performs daily. ‡
Q. What are these motions called ?
A. The first is called its annual motion, and the last its diurnal.
Q. What is caused by the annual motion ?
A. The change and variety of the seasons.
Q. What is caused by the diurnal motion ?
A. The succession of day and night.

* Being flattened at the Poles.
† The circumference of the earth is 24,872 miles, and its diameter is 7,914 miles.
‡ There is also, a third motion of the earth, called the PRECESSION OF THE EQUINOXES, which is a slow motion of the two points, where the equator cuts the ecliptic, which are found to move backward and forward fifty seconds every year.

CHAPTER II.

Of the Surface of the Earth—Land.

Q. Of what does the earth consist ?

A. Land and water.

Q. What are the great divisions of the earth ?

A. The earth is divided into five great portions, namely, Europe, Asia, Africa, America, & Australia.*

Q. What are the smaller divisions of the earth ?

A. The smaller divisions are continents, islands, peninsulas, isthmuses, promontories, &c.

Q. What is a continent ?

A. A continent is a large tract of land, containing many nations, as Europe.

Q. What is an island ?

A. An island is a smaller tract of land, entirely surrounded by water, as Great Britain.

Q. What is a peninsula ?

A. A peninsula is a tract of land, almost surrounded by water, as the Morea in Greece.

Q. What is an isthmus ?

A. An isthmus is a narrow neck of land which joins a peninsula to a continent, or two continents together, as the isthmus of Corinth, which joins the Morea to the Continent, and the isthmus of Suez, which joins Africa to Asia.

Q. What is a promontory ?

A. A promontory is a point of land stretching out into the sea, the end of which is called a cape, as the Cape of Good Hope in Africa.

Water.

Q. Of what does the water consist ?

A. The water consists of oceans, seas, gulfs, straits, rivers, and lakes.

---

* Until very lately it was customary to say that the earth was divided into four quarters; but the term Australia (or, as it is sometimes written, Australasia) has been adopted by modern geographers to denote the numerous islands in the great Pacific Ocean, which by some are classed under two names, viz, Australasia and Polynesia.

Q. What is the Ocean ?

A· The Ocean is a vast collection of water which encompasses the earth, and is subdivided into the Pacific, Atlantic, Indian, Southern, & Northern Oceans.

Q. What is a sea ?

A. A sea is a smaller collection of water than an ocean, as the Mediterranean, the Baltic, the Black Sea, &c.

-Q. What is a gulf ?

A. A gulf or bay is a part of the sea running a considerable way inland, as the Gulf of Mexico, the Gulf of Guinea, the Bay of Biscay, &c.

Q. What is meant by a strait ?

A. A strait is a narrow part of the sea joining one sea to another, as the Strait of Gibraltar, and the Strait of Dover.

Q. What is a river ?

A. A river is a stream of water rising in the land and flowing into the sea, as the Thames.

Q. What is a lake ?

A. A lake is a large body of water, surrounded by land, as the lake of Geneva, Lake of Ontario, &c.

CHAPTER III.

Of Europe.

Q. How is Europe bounded ?

A. Europe has the Frozen Ocean to the North, Asia, to the East, the Mediterranean Sea to the South, and the Atlantic Ocean to the West ?

Q. What is the size of Europe ?

A. Europe contains five millions of square miles, being the smallest of the great divisions of the earth.

Q. For what is Europe celebrated ?

A. Europe is celebrated for the learning. politeness activity, and intelligence of its inhabitants, as well as for the fertility of its soil, and the temperature of its climate.

Q. What are the principal nations of Europe ?

A. Europe comprehends the following independent countries, viz : Sweden, Russia, Denmark,

Prussia, Germany, Hanover, Saxony, Wirtemberg, Bavaria, Austria, Turkey, Greece, France, the Netherlands, Switzerland, Parma, Naples, Rome, Sardinia, Tuscany, Spain, Portugal, Great Britain and Ireland, and the republic of the Seven Islands.

CHAPTER IV.

Of Sweden, Norway and Lapland.

Q. How is Sweden bounded?

A. Sweden is bounded on the South by the Baltic, on the East by the Gulf of Bothnia, on the West by Norway, and on the North by the North Sea.

Q. What kind of a country is Sweden with respect to size and climate?

A. Sweden is a very large country, excessively cold in winter, having many parts of its mountains always covered with snow; but the air is pure and wholesome.

Q. How long does the winter continue in Sweden?

A. Seven or eight months, and in some parts nine; but the summer comes on so suddenly that the valleys are green in a few days.

Q. What kind of soil has Sweden?

A. The soil of Sweden in general is rather barren than fertile, the country being full of mountains, rocks, woods, and lakes

Q. What are the principal commodities of Sweden

A. It has many mines of silver, copper, lead, and iron, also, vast quantities of pitch, tar, masts, deals, hides, buckskins, fur, tallow, and honey, which the Swedes export to foreign countries.

Q. What is the character of the Swedes?

A. The Swedes are of a robust constitution, and able to sustain the hardest labour.

Q. What are the manners of the Swedes?

A. They are polished and warlike, brave, active, friendly to science, and luxurious.

Q. What is the religion of Sweden?

A. The Lutheran.

Q. What is the capital of Sweden?

A. Stockholm.

Q How far is Stockholm from London?

A. Stockholm is distant from London 895 miles.

Q. How many inhabitants does Sweden contain?

A. Sweden contains three millions of inhabitants.

Q. What was the ancient name of Norway and Sweden?

A. Scandinavia. Lapland was unknown to the ancients.

Norway

Q. What is Norway?

A. Norway is a very mountainous and extensive country in the North of Europe, but with a small population not exceeding one million.

Q. To whom does Norway belong?

A. Norway formerly belonged to Denmark, but it is now subject to Sweden.

Q. What are the chief towns of Norway?

A. Christiana and Bergen, the former containing 10,000 inhabitants, and the latter 19,000.

Q. In what does the wealth of Norway consist?

A. The chief wealth of Norway arises from its timber, and iron and copper mines, with which it supplies the greater part of Europe.

Q. What is the religion of Norway?

A. The religion of Norway, like that of Sweden, is the Lutheran; and the language is a dialect of the Gothic.

Q. what is the climate of Norway?

A. It is in general very damp, much exposed to heat during its short summer, and excessively cold in winter; yet for the most part it is considered a very healthy country.

Q. What is the length of the longest day in Norway?

A. The longest day in the north is two months, and in the south eighteen hours only; the difference the latitude of this country being so great as to cause this variation.

Q. What is the character of the Norwegians?

A. The Norwegians are a strong, brave, and friendly people, but passionate and revengeful.

Lapland

Q. What is Lapland.

A. Lapland is the most northern country in Europe, and is divided into Danish, or north Lapland ;† Swedish, or South Lapland ; and Russian, or East Lapland.

Q. To whom does Lapland belong?

A. The greatest part of Lapland belongs to Sweden, and the remaining part to Russia. The precise boundaries of Lapland are unknown.

Q. What are the chief towns of Lapland?

A. The principal towns of Lapland are *Wardhuus* in Danish Lapland, and *Kola* in Russian Lapland.

Q. What kind of a country is Lapland?

A. It is extremely cold and barren, and consists chiefly of mountains, fens, forests, and lakes, which are covered with snow two-thirds of the year. Its forests consist principally of fir, and its pastures are full of rein-deer

Q. Describe the Laplanders.

A. In consequence of the very rigorous climate of Lapland its inhabitants are a diminutive race, seldom exceeding four feet in height. Their features are very coarse, and their complexion swarthy. Their language is rude and scarcely articulate.

Q. What is the population of Lapland?

A. It is very thin and scattered, and the whole does not exceed 60,000.

CHAPTER VI.
Of Russia in Europe.

Q. What was the ancient name of Russia?

A. Russia was anciently divided into two great parts; namely, Sarmatia and Scythia; the former situated to the West, and the latter to the East. It

---

* That is to say the sun does not set for two months.

† Danish Lapland is sometimes called Finmark. This part of Lapland was ceded to Sweden, together with Norway.

has also been named Muscovy from Moscow, its ancient capital.

Q. How is the Russian Empire divided?

A. The Russian Empire is divided into Russia Proper, Lapland, Finland, Poland and Asiatic Russia.

Q. What is Finland?

A. Finland,* formerly a province of Sweden, was invaded by the Russians in the year 1308, and added to their empire in 1809, together with the adjacent islands and part of Lapland. Its chief town is *Abo*.

Q. What is Russian Lapland?

A. Russian Lapland† is the eastern part, situated north of Finland, and is included in the government of Archangel; but the inhabitants trade chiefly with the Swedes and Norwegians.

Q. What is Russia proper?

A. Russia Proper, situated in Europe, is bounded E. by Siberia, S. by Poland, W. by Finland, and N. by the Ocean. This division is by far the most populous.

Q. What is Polish Russia, or Poland?

A. Before the decision of the congress of Vienna in 1814 this country had long lost its independence, and was divided between Russia, Austria, Prussia, and France; but at the overthrow of Bonaparte that portion, which belonged to Russia ‡, was again erected into a kingdom under the protection of the Emperor of Russia, who is styled its king.

Q. Is not the Russian empire of great extent?

A. Yes; it is the largest country in the world, comprehending all the northern part of Asia, and the east of Europe.

Q. What kind of climate has Russia?

A. As the extent of Russia is very great, there is a great variety of climate; in some parts the winter is very severe, and in others very mild.

* Finland was part of the ancient Scandinavia, then called Finigia, the people Finn.—*Tacitus*.
† Now including part of Swedish Lapland.
‡ About two-thirds of the ancient kingdom.

Q. What is the appearance of the country ?

A. The gen r l f   of h country is marshy. full of extensive plains, forests, lakes, and rivers. The north rn provinces for the most part are barren and but little cultivated ; but the middle and southern provinces are very fertile and yield plenty of corn.

Q. What is the produce of Russia ?

A. Russia produces large quantities of cotton and silk, skins furs, leather, tallow, hemp, honey, and wax.

Q. What is the general character of the Russians ?

A. The Russians are in general robust, well shaped and of a fair complexion. They were formerly the most ignorant, and even the most savage people in the world ; but they are now making very great progress in every elegant improvement and refinement.

Q. What is the metropolis of Russia ?

A. St. Petersburgh, built by Peter the Great.

Q. How far is St. Petersburgh from London ?

A. One thousand two hundred and sixty-five miles.

Q. What is the population of Russia ?

A. Thirty-six millions in Europe, and five in Asia.

Q. What are the principal rivers in Russia ?

A. The principal rivers in Russia are the Wolga, the Don, the Neva, the Niester, the Dwina, and the Vistula. The Wolga is 3000 miles in length.

CHAPTER VII.

Of Poland.

Q. How is Poland bounded ?

A. Poland is bound N. and E. by Russia, W. by Prussia and Germany, and S. by Turkey and Hungary.

Q. What is the face of the country ?

A. Poland in general is level, and well watered with lakes and rivers, and abounds with rich pastures

Q. What kind of climate has Poland ?

A. The climate of Poland is in general temperate and healthy ; and, being situated nearly in the middle of a large continent, the weather is less changeable than in those countries which border on the Ocean.

Q. What are the chief products of Poland ?

A. They are leather, hemp, flax, honey, and wax, besides large mines of salt.

Q. What are the chief cities of Poland ?

A. The chief cities of Poland are Cracow, its former capital, and Warsaw, its present.

Q. What is the character of the Poles?

A. They are in general handsome, tall, and well made ; robust, hardy, and brave ; but must be censured as being meanly submissive to their superiors, arrogant to their equals, and tyrannical to their inferiors.

Q. What is the population of Poland, and what its prevailing religion ?

A. The population of Poland is supposed to be about thirteen millions and a half, and its prevailing religion the Roman Catholic.

CHAPTER VIII.

Of Denmark.

Q. What kind of a country is Denmark ?

A. Denmark is a very small country, in general flat, and the soil very fruitful, producing plenty of corn and vegetables. The climate is very cold and severe.

Q. What compose the Danish dominions?

A. The Danish Dominions consist of a peninsula, containing Jutland, Sleswick, and Holstein, and of the islands of Zealand, Funen, Laland, Falster ; and Iceland and Faro, with others of less note.

Q. What is the character of the Danes?

A. They are a brave and warlike people, very frugal and industrious.

Q. What is the metropolis of Denmark?

A. Copenhagen in the Island of Zealand.

Q. How far is Copenhagen from London ?

A. Five hundred and ninety-five miles.

Q. What is the population of Denmark ?

A. Three millions.

Q. What are the government and religion of Denmark ?

A. The government is an absolute monarchy, and the religion, Lutheran.

Q. What was the ancient name of Denmark, or rather Denmark Proper?

A. Cimbrica.* The people were called *Cimbri* by the Romans, and *Cimmerii* by the Greeks.

Q. What is Denmark Proper ?

A North and South Jutland†, down as far as Holstein.‡ South Jutland is usually called Sleswick.

Q. Whence is the origin of the *Cimbri ?*

A. History does not particularly inform us ; but it is generally conceived they were a colony of Scythians from the country between the Caspian and Euxine Seas.

Q. When were the inhabitants of Cimbrica first called Danes?

A. In the seventh century, and they were so called from the *Dani,* a distinguished people from Norway, who conquered Cimbrica, in the period above mentioned.

CHAPTER IX.

Of Prussia.

Q. What kind of country is Prussia?

A. Prussia is a very rich and fertile country, producing a great deal of flax, hemp, and corn.

---

* It was usually called by the Greeks *Cimbrica Chersonessus,* or the Peninsula of Cimbrica.

† This country received the name of Jutland from a people called the Jutes, who once inhabited part of it. It is to be observed that Cimbrica, or the Danish nation, in its early periods, consisted of distinct tribes, or independent bodies of adventurers, without laws or any regular form of government. The Cimbri and Teutons are among the first people of Europe mentioned in history ; the latter principally reside in the islands of Zealand and Funen.

‡ Holstein is considered as a part of the present and ancient Germany.

Q. What composes the kingdom of Prussia ?

A. The Prussian dominions are situated in three different countries; namely, in Germany, Prussia and Poland. Some of the provinces in Germany, situated on the Rhine, are considerably detached.

Q. What was the ancient name of Prussia ?

A. The ancient name of Prussia is involved in obscurity. Its original inhabitants appear to have been a bold and warlike people, descended from the *Sclavonians.* Prussia Proper, or Ducal Prussia, was formerly subject to the Teutonic knights, who conquered it in the thirteenth century. It was raised to a dukedom in 1525, and formed into a kingdom in the year 1701.

Q. What is the character of the inhabitants ?

A. They are of a strong constitution, laborious and brave.

Q. Which is the chief city of Prussia ?

A. Berlin is its capital.

Q. How far is Berlin from London ?

A. Five hundred and ninety miles.

Q. What is the population of Prussia ?

A. Before the late war with France, Prussia contained 8,000,000 of inhabitants : during that contest the number was greatly diminished ; but eventually she gained 2,000,000 of subjects ; and the king, in addition to his former titles, received that of the duke of Saxony, with a large portion of that country.

Q. What is the prevailing religion of Prussia ?

A. It is the Protestant, which is divided into Lutheran and Calvinistic ; but the former has the ascendancy ; all other religions are tolerated.

Q. What is the government of Prussia ?

A. An absolute monarchy, and the succession is hereditary.

CHAPTER X.

Of the Kingdom of the Netherlands: or the States of Holland and Belgium.

HOLLAND.

Q. What is meant by the Netherlands?

A. The Netherlands, or Low Countries, so called by Charles V. of Spain, and Germany, from their low situation, comprised Holland and Belgium: the former is sometimes called Batavia, and the latter the Catholic Netherland.*

Q. What does Holland comprehend?

A. Under the name of Holland, are comprehended the Seven United Provinces of the Netherlands, of which that province is the principal.

Q. How is Holland, or the Seven United provinces bounded?

A. Holland is bounded N. and W. by the German Ocean, S. by Belgium. and E. by Germany.

Q. When did Holland separate itself from the other parts of the Netherlands, and what occasioned it?

A. The division of Holland from Belgium was occasioned by the tyranny of its sovereign, Philip II., king of Spain, which caused an insurrection of the inhabitants in 1609, when the Spaniards were compelled to declare the Hollanders a free people.†

---

*The name Catholic, was applied to this division from its remaining subject to the Crown of Spain, after the revolt of the other Provinces from their Catholic prince Philip.

† These people were immediately afterwards acknowledged by all Europe to be an independent state, under the title of the Seven United Provinces; but Belgium or the Catholic Provinces remained subject to the Spanish government till the year 1700, when they were ceded to the house of Austria; since which they have been dismembered by the Prussians, Dutch, &c.; conquered by the French, united with Holland; and lastly, in the year 1830, formed into a separate kingdom.

Q. What kind of country is Holland?

A. It is a very small country, but more populous than any other in the world of so small an extent. The land is everywhere lower than the sea, which is kept out by means of dykes.

Q. What is the character of the inhabitants: and by what name are they generally called?

A. They are generally called Dutch, and are rather ceremonious than polite ; but are plain and industrious.

Q. What is the occupation of the Dutch?

A. Trade and getting of money, for which their nation is proverbial.

Q. What is the name of their chief city?

A. Amsterdam.

Q. How far is Amsterdam from London?

A. Two hundred and six miles.

Q. What is the population of Amsterdam?

A. Two hundred thousand.

BELGIUM.

Q. What kind of country is Belgium?

A. The air of Belgium is temperate, and its soil extremely fertile in corn and pasture ; and there are several large manufactories of lace, lawn, &c.

Q. How is Belgium bounded?

A. Belgium is bounded N. by Holland, E. by Germany, S. by France, and W. by the German Ocean. This country is very populous.

Q. Why are the Catholic Netherlands sometimes called Belgium, and at other times Flanders?

A. They received the name of Belgium from the Romans in the time of Julius Cæsar, and that of Flanders for its principal division, which is so named.

Q. By what name are the inhabitants of Belgium generally known, and for what are they celebrated?

A. The inhabitants of Belgium are sometimes called Belgians, but they generally go by the name of Flemings from Flanders, and are celebrated for their invention of the art of staining glass : also for making cambrics, laces, &c., and have had the honour of forming a school for painting, which vied with that of Italy.

Q. What are the chief cities of Belgium ?

A. They are Antwerp, Brussels, and Ghent ; but the principal of them is Brussels, The population of Brussels is about 80,000 ; Ghent 60,000 ; and Antwerp, 50,000.

Q. What is the population of the Netherlands ?.

A. It is about six millions.

Q. What was the ancient name of the Netherlands ?

A. Belgica, which was one of the principal divisions of ancient Gaul.* The people were called Belgæ.

Q. What was the ancient name of Holland ?

A. Batavia, so called from the Batavi, the principal people of this country. The Batavi were a branch of the Batti, from Germany.

CHAPTER XI.

Of Germany.

Q. What kind of country is Germany ?

A. Germany is a very large country, situated in the middle of Europe, and is remarkable for being divided into a great number of independent states, which are politically united for the maintenance of the external and internal inviolability of Germany, and form what is called the Germanic Confederation.

Q. Are these independent states of equal magnitude and importance ?

A. No ; some are kingdoms, the rest are grand duchies, marquisates, principalities, bishoprics, or free cities.

Q. Which are kingdoms ?

* In the time of Julius Cæsar, Gaul was divided into three great nations ; namely, the eltæ, the Acquitani, and the Belgæ. Of these the Celtæ or Celts, were the most ancient and numerous. The words Gallia is the Latinized term of Gaul, which was the name given to this country by the Romans.

A. The kingdoms are Austria, Prussia, Bavaria, Saxony, Hanover, and Wirtemberg.

Q. Which are the free cities, and how are they governed?

A. The free cities are Lubec, Frankfort, Bremen, and Hamburgh, which are sovereign states in themselves, governed by their own Magistrates.

Q. What is the climate of Germany?

A. The air is temperate and wholesome, but more inclinable to cold than heat.

Q. What is the soil of Germany?

A. It is very rich and fertile, both for corn and pasture.

Q. What is the character of the Germans?

A. They are in general open and free, good soldiers, inured to labour, quick in the sciences, dexterous in manufactures, and fruitful in inventions.

Q. What is the population of Germany, exclusive of Austria and Prussia?

A. Twenty-five millions.

Q. What are the principal rivers of Germany?

A. The principal rivers of Germany are the Danube, Rhine, Maine, Weser, Elbe, and Oder.

Q. What was the ancient name of Germany?

A. Germania, the limits of which were very different from those of modern Germany.

Bavaria, Saxony, Hanover, Wirtemberg, Baden, and Mecklenburgh.

Q. Describe Bavaria?

A. The kingdom of Bavaria is situated in the south of Germany, bordering on Italy and Switzerland, with the kingdoms of Wirtemberg on the west, Saxony on the north, and Austria on the east.

Q. What is the capital of Bavaria?

A. It is Munich, one of the most pleasant cities in Germany. The houses are high, and the streets spacious, and intersected with canals. It is situated on the Iser, and contains about forty thousand inhabitants. The population of Bavaria is about four millions.

Q. Describe Saxony.

A. The kingdom of Saxony, so celebrated in the history of Europe, comprises part of the former electorate of Saxony, and of Upper and Lower Lusatia, with the duchies of Gotha and Weimar. The number of its inhabitants is about three millions.

Q. Which are the chief cities?

A. Dresden and Leipsic. The former, situated on the Elbe, is the capital of the kingdom, and is noted for its handsome, appearance; the number of its inhabitants is about sixty thousand. Leipsic is famous for its great fairs, and for the celebrated battle of 1814 between Buonaparte and the Allied Powers of Austria, Russia, and Prussia.

Q. How far is Dresden from London?

'A. Five hundred and ninety miles.

Q. Describe Hanover.

A. The kingdom of Hanover, situated on the north of Germany, is composed of the duchies of Luneburg, Bremen, Verden, and Saxe Lauenburg, with the countries ot Calenburg, Grubenhagen, &c. Its capital is Hanover, containing about fifteen thousand inhabitants, and is noted for the burial place of Zimmerman, and for having given birth to Dr. Herschel. Hanover contains about three millions of inhabitants.

Q. Describe Wirtemberg?

A. The kingdom of Wirtemberg, situated in the south of Germany, is the smallest of the recent German kingdoms, but one of the most fertile. Its population is about two millions. The capital is Stutgard, containing about 20,000 inhabitants.

Q. What is Baden?

A. The Grand Duchy of Baden is one of the smallest of the minor states of Germany. It is situated between the Rhine and the kingdom of Wirtemberg. It is very fertile in wine, corn and fruit, and has superfluity of fish and wood. The capital is Carlsruhe.

Q. What is Mecklenburgh?

A. Mecklenburgh is a small principality in the

north of Germany, whose house has been frequently
divided.* At present there are two lines, Schwerin
and Strelitz. It is from the latter that her majesty
Charlotte, queen of George the Third, of Great Britain, was descended.

CHAPTER XII.
Of Austria.

Q. What countries compose the Austrian Empire ?

A. The Empire of Austria comprehends Austria,
Bohemia, Hungary, and part of Poland,† Mora-
via, Transylvania Sclavonia,'Croatia, the Tyrol, parts
of Dalmatia, and Lombardy, including the territories
of Milan and Venice.

Q. What is the general character of the Austrians ?

A. They are sensible, polite, well versed in the
arts and sciences, and warlik.

Q. What is the character of the Hungarians ?

A. They are well made, brave, hardy and military ;
but haughty, proud, and vindictive.

Q. What is the appearance of the country ?

A. It is in most of the regions subject to Austria,
mountainous, and in many parts covered with vast forests. The soil would have been productive, had not
the ravages of war prevented its proper cultivation.

Q. What is the capital of Austria ?

A. Vienna.

Q How far is Vienna from London ?

A. Seven hundred and thirty miles.

Q. How many inhabitants are there in the Austrian dominionns ?

A. Twenty-eight millions. ‡

* The princes of Mecklenburgh are descended from the ancient
Vandals who inhabited the country.

By the partition of Poland in 1793 Austria acquired about
one-sixth of that country and above 4,000,000 of its inhabitants.
In the late wars with France she lost Belgium and Flanders ;
but on the other hand acquired the Venetian territories, so that
she is still a very powerful empire, having a population of about
28,000,000 of inhabitants.

‡ Owing to the increase of population since the census of Austria, and that of newly acquired territories, the inhabitants of
this empire are rated as above.

CHAPTER XIII.

Of Turkey

Q. What kind of country is Turkey ?

A. Turkey is a very large empire, having a part in Eurpoe, a part in Asia, and a part in Africa.

Q. What was the ancient name of Turkey in Euroje ?

A. The greater part was anciently called *Graecia** or Greece, which included Macedonia, Thracia, Thessalia, Peloponnesus, &c.

Q. What is the origin of the Turks ?

A. The Turks derive their origin from the Avares, a tribe of Huns, who dwelt in Great Tartary till forced by the Hun, of the South to abandon their country, when a considerable colony seized upon Turkey, and founded this empire.

Q. What is the climate of Turkey ?

A. It is in general wholesome and pleasant; but the inhabitants are frequently visited with the plague, which is partly attributed to their indolence and to their want of cleanliness.

Q. What is the character of the Turks?

A. They are a strong, well made people, but dislike labour, and derive no advantage from the fine soil they inhabit. They are ignorant, but frugal, temperate, and charitable.

Q. What is the capital of the Turkish Empire ?

A. Constantinople.

Q. How far is Constantinople from London ?

A. One thousand five hundred miles.

---

* Greece Proper was a very inconsiderable spot in Europe, yet more formidable than all others in war, and more celebrated in the arts of peace. The inhabitants called it *Hellas*, and themselves *hell-nes*; the Romans called them· *Graeci*. After the subjugation of their country by the Turks, they remained in a state of vassalage under their barbarous conquerors; but after a desperate and sanguinary war, they succeeded in their independence. Greece forms now a separate kingdom, of which Otho, a son of the king of Bavaria, is sovereign.

Q. What is the population of Turkey ?

A. Turkey in Europe has eight millions, in Asia, ten millions, and in Africa, two millions five hundred thousand.

Q. What are the principal mountains of Turkey in Europe ?

A. The principal mountains in European Turkey are Athos, Pindus, Olympus, Parnassus, and Haemus.

Q. What are the principal rivers ?

A. The Danube, the Sava, the Pruth, and the Don.

CHAPTER XIV.

Of France.

Q. What kind of country is France ?

A. France is a very large and populous country, containing about thirty-three and a half millions of inhabitants, seven hundred and eighty cities, and forty-one thousand towns and villages.

Q. What are its soil, its climate, and its produce ?

A. The air is pure and wholesome ; and the soil, which is agreeably diversified, produces all the necessaries of life, and among its luxuries, some excellent wines.

Q. What is the character of the French ?

A. The French are a polished people, gallant and courageous, but light, inconstant, and excessively vain.

Q. Of what are they particularly fond ?

A. Of the arts and sciences, and of games, exhibitions and dancing.

Q. What is the capital of France ?

A. Paris on the river Sein.

Q. How far is Paris from London ?

A. Two hundred and ten miles.

Q. What are the principal cities of France ?

A. Rouen, Tours, Lyons, Nismes, Montpelier, and Toulouse.

Q. What are its chief seaports ?

A. Havre, Brest, Nantes, Bordeaux, Bayonne, Marseilles, and Toulon.

Q. What are the principal rivers?

A. The Sein, the Loire, the Garonne or Gironde, and the Rhone.

Q. What is the religion of France?

A. The religion established by law is the Roman Catholic, but all others are tolerated.

Q. What are the foreign possessions belonging to France?

A. The isles of Martinique and Guadaloupe in the West Indies, Cayenne in South America, Pondicherry in the East Indies, the Isles of Bourbon in the Indian Ocean, and Algiers, Senegal, and the Isle of Goree, in Africa.

Q. What was the ancient name of France?

A. Gallia or Gaul. The people were called *Galli* by the Romans, *Galatae* by the Greeks, and *Celtae* by themselves. It took its present name from the Franks, a people formerly inhabiting Germany, who afterwards conquered part of Gaul.

CHAPTER XV.

Of Switzerland.

Q. What kind of country is Switzerland?

A. No country affords a greater variety of surface; sublime mountains, frightful precipices, regions of snow that never melts, glaciers that resemble seas of ice, fertile valleys, cottages, and hanging vines diversify the scene.

Q. What is the nature of the climate?

A. It is wholesome and agreeable; but the winter frosts are extremely severe, and many of the mountains are always covered with snow.

Q. For what are the Swiss remarkable?

A. For their simplicity of manners, purity of morals, and strong attachment to their native country.

Q. For what are the men particularly distinguished?

A. For their strength, courage, and sobriety.

Q. In what do the women excel?
A. The women excel in modesty, and in the maternal and domestic duties.
Q. What are the chief towns in Switzerland?
A. Berne, its capital, and Geneva and Lausanne.
Q. How far is Berne from London?
A. Four hundred and forty-five miles.
Q. What is the population of Switzerland?
A. Two millions.
Q. What was the ancient name of Switzerland?
A. Helvetia. The people were called *Helvetii*, and were famed for their bravery.

CHAPTER XVI.

Of Italy.

Q. What kind of country, is Italy?
A. It is the most celebrated country in .Europe, having been formerly the seat of the Roman empire, and being still the residence of the Pope. It is so fine and fruitful a country that it is often called the garden of Europe.
Q. Describe its climate, soil, and productions?
A. The air is in general temperate and wholesome; and the soil very fertile, producing a great variety of wines, and the best oil in Europe; also, wheat, rice, oranges, citrons, &c., and all sorts of fruits, honey, and silk, with sugar and cotton.
Q. What is the character of the Italians?
A. They are polite, sensible, and fond of the arts. They excel in sculpture, painting, and music; but are revengeful and superstitious.
Q. How is Italy bounded?
A. It is bounded on the North by the Alps; on the West by France and the Mediterranean; on the South by the Mediterranean; and on the East by the Adriatic.
Q. Into how many kingdoms and states is Italy at present divided?
A. Into eight, viz.: 1st. The kingdom of Sardinia,

comprising the island of that name; and Savoy, Pied-
mont, and Genoa ; 2nd. Venice and Milan, which
belongs to the Emperor of Austria; 3rd. Parma ; 4th.
Modena and Massa ; 5th. Lucca ; 6th. Tuscany ; 7th.
The States of the Church of Rome, which are gov-
erned by the Pope ; 8th. The kingdom of the two
Sicilies, comprising the island of that name, and
Naples.

Q. Which is the most celebrated city of Italy ?
A. Rome.
Q. How .far is Rome from London ?
A. Eight hundred and sixty-three miles.
Q. What are the other chief cities of Italy ?
A. They are Milan, Venice, Naples, Turin, Genoa
and Florence.
Q. What are its chief mountains ?
A. The Alps, which divide it from France and
Switzerland, the Appenines, and Vesuvius, a volcano.
Q. What are the chief rivers ?
A. They are the Po, the Adige, the Arno and the
Tiber.
Q. What is the population of Italy ?
A. About twenty-two millions.
Q. What was the ancient name of Italy ?
A. *Italia*, one of the noblest countries in Europe,
not only in climate, but in importance.

CHAPTER XVII.

Of Portugal.

Q. What is Portugal ?
A. Portugal is a small kingdom, containing about
three and a half millions of inhabitants ; its capital is
Lisbon.
Q. Describe the climate of Portugal ?
A. It is pure and wholesome, the heat being mod-
erated by frequent westerly winds.
Q. What kind of soil has Portugal ?
A. There are many barren mountains, but some
parts are rich and fertile, producing excellent wines
and fruit of all kinds.

Q. What is the character of the Portuguese?

A. They are in general brave, austere, and superstitious; but sensible and well informed in the arts, sciences, and trade.

Q. How far is Lisbon from London?

A. One thousand and ten miles.

Q. What is the chief river?

A. The Tagus on which is Lisbon.

Q. What was the ancient name of Portugal?

A. Lusitania: the people were called *Lusitani*.

CHAPTER XVIII.

Of Spain.

Q. What kind of country is Spain?

A. Spain is a large country about the size of France. Its population is about 13,000,000. Madrid is the capital city.

Q. What is the climate of Spain?

A. The air is generally very hot, which frequently obliges the inhabitants to lie down after dinner, and sit up late at night.

Q. What is the character of the Spaniards?

A. They are proud and haughty, grave and polite, and attached to their country, of which their patriotic struggle against Bonaparte's usurpation affords a noble specimen.

Q. For what are the Spaniards celebrated?

A. For their sobriety and probity; still their idleness makes them a miserable people, though they live in one of the finest countries in the world.

Q. What is the soil of Spain?

A. Spain in general is vey fertile, but is not cultivated through the pride and laziness of its inhabitants. Many parts are overrun with woods, and encumbered with lofty mountains.

Q. What are the most remarkable cities?

A. Saragossa, Salamanca, Toledo, Cordova, Grenada, Seville, and Badajoz.

B

Q. What are its chief seaports ?

A. Corunna, Ferrol, Cadiz, Gibraltar, and Barcelona

Q. What is the produce of Spain ?

A. Wines, fruits of various sorts, wool, honey, cork, anchovies, &c.

Q. How far is Madrid from London ?

A. Eight hundred miles.

CHAPTER XIX.

Of Great Britain.

Q. What constitutes the British Empire ?

A. The British Empire consists of Great Britain and Ireland with some adjacent isles of Europe. This Empire was formerlydivided into three distinct kingdoms, namely, England, Scotland, and Ireland ; but it is now united under one government.*

Q. What kind of country is Great Britain ?

A. It is a large island, advantageously situated for commerce. The southern part (or England) is very fertile, and the land in a high state of cultivation, abounding with every production necessary for the sustenance of man.

Q. What is the character of the inhabitants ?

A. They are a brave and polished people ; honest, industrious, and hospitable ; fond of liberty, and at-

---

* The foreign possessions of England, which are situated in every quarter of the globe, are very extensive.

Those in ASIA consist of a vast territory in the East Indies, containing more than 30,000,000 of inhabitants. In AFRICA, the Cape of Good Hope, St Helena, Sierra Leone, with several forts on the Coast of Guinea. In AMERICA, Canada, Labrador, New Brunswick, Nova Scotia, the countries around Hudson's Bay, Newfoundland, &c., with many valuable islands in the West Indies. In EUROPE, the fortress of Gibraltar in Spain, the Ionian Isles and the Island of Malta in the Mediterranean Sea. In AUSTRALASIA, various Colonies in New Holland ; with some other possessions of less note.

tached to their country. They are likewise great patrons of the arts, and encouragers of literature.

Q. What are the ancient names of England, Ireland, and Scotland?

A Of England, *Britannia** ; Ireland, *Hibernia* † ; Scotland, *Caledonia.* ‡

Q. How is England divided?

A. Into forty counties or shires, which are as follow:

Six northern,	Three south-eastern,
Four bordering on Wales,	Four southern,
Twelve midland,	Three south-western.
Eight eastern,	

Q. Name the six northern counties with their chief towns.

Counties.	*Chief Towns.*
A. Northumberland,	Newcastle,
Cumberland,	Carlisle.
Durham,	Durham.
Yorkshire,	York.
Westmoreland,	Appleby.
Lancashire, or Lancs,	Lancaster.

Q. Which are the four bordering on Wales with their chief towns?

A. Cheshire,	Chester.
Shropshire,	Shrewsbury.
Herefordshire.	Hereford.
Monmouthshire	Monmouth.

Q. Which are the twelve midland counties with their principal towns?

Counties.	*Chief Towns.*
A. Nottinghamshire, or Notts.	Nottingham.
Derbyshire,	Derby.
Staffordshire,	Stafford.

* So called by the Romans. The original Celtic name was *Bretton,* signifying a high mountainous country. Pliny says its ancient name was *Albion* ; but that *Britannia,* a name of the same meaning, became the prevalent appellation.

† So called by Cæsar, Tacitus, and Pliny ; but named *Ierns* by Strabo.

‡ Tacitus.

Leicestershire,	Leicester.
Rutlandshire,	Oakham.
Northamptonshire,	Northampton.
Warwickshire,	Warwick.
Worcestershire,	Worcester.
Gloucestershire,	Gloucester.
Oxfordshire,	Oxford.
Buckinghamshire, or Bucks,	Buckingham.
Bedfordshire, or Beds,	Bedford.

Q. Which are the eight eastern counties with their chief towns?

A. Lincolnshire,	Lincoln.
Huntingdonshire,	Huntingdon.
Cambridgeshire,	Cambridge.
Norfolk,	Norwich.
Suffolk,	Ipswich.
Essex,	Chelmsford.
Hertfordshire, or Herts.	Hertford.
Middlesex;	London.

Q. Which are the three South-eastern counties with their chief towns?

A. Surrey,	Guildford.
Kent,	Canterbury.
Sussex.	Chichester.

Q. Which are the four southern counties with their chief towns?

A. Berkshire, or Berks,	Reading.
Wiltshire, or Wilts,	Salisbury.
Hampshire, or Hants,	Winchester.
Dorsetshire,	Dorchester.

Q. Which are the three south-western counties with their chief towns?

A. Somersetshire,	Bath.
Devonshire,	Exeter.
Cornwall,	Launceston.

Q. What is the population of England and Wales?

A. In 1841 the population, including the Army and Navy, was nearly sixteen millions.

Q. What are the chief rivers in England?

A. The Thames, the Severn, the Medway, the Humber, the Mersey, and the Trent.

Q. What are its chief mountains?

A. The Peak in Derbyshire ; the Cheviot hills between England and Scotland ; the Catswold in Gloucestershire ; and the Mendip in Somersetshire.

Q. What are the chief ports ?

A. Newcastle, Sunderland, Hull, Yarmouth, London, Dover, Deal, Southampton, Plymouth, Falmouth, Bristol, and Liverpool.

Q. What are the principal dock-yards ?

A. Chatham, Portsmouth, and Devonport.

Q. What is the metropolis of England ?

A. London, one of the largest cities in the world and of the most extensive commerce.

CHAPTER XX.

Of Wales.

Q. Into how many counties is Wales divided ?

A. Into twelve : six northern, and six southern.

Q. Which are the six northern counties with their chief towns ?

Counties.	Chief Towns
A. Flintshire,	St. Asaph.
Denbighshire,	Denbigh.
Carnarvonshire,	Carnarvon.
Isle of Anglesea,	Beaumaris.
Merionethshire,	Dolgelly.
Montgomeryshire,	Montgomery.

Q. What are the six southern counties with their chief towns ?

A. Radnorshire,	Radnor.
Cardiganshire,	Cardigan.
Pembrokeshire,	Pembroke,
Carmarthenshire,	Carmarthen.
Brecknockshire,	Brecknock.
Glamorganshire,	Cardiff.

Q. Which are the chief mountains in Wales ?

A. The chief Welsh mountains are Snowdon, Cader-Idris, and Plinlimmon.

CHAPTER XXI.

Of Scotland.

Q. How is Scotland divided?

A. Into two parts, one north, or the Highlands, and one south, or the Lowlands. The whole contains thirty-three counties.

Q. Which are the first six counties with their chief towns?

Counties.	Chief Towns.
A. Edinburgh,	Edinburgh.
Haddington,	Haddington.
Roxburgh,	Jedburgh.
Selkirk,	Selkirk.
Peebles,	Peebles.
Lanark,	Glasgow.

Q. Which are the second six counties with their chief towns?

A. Dumfries,	Dumfries.
Wigton,	Wigton.
Kircudbright,	Kircudbright.
Ayr,	Ayr.
Dumbarton,	Dumbarton.
Bute,	Rothsay.

Q. What are the names of the third six counties with their chief towns?

A. Caithness,	Wick.
Renfrew,	Paisley,
Stirling,	Stirling.
Linlithgow,	Linlithgow.
Argyle,	Inverary.
Perth,	Perth.

Q. What are the names of the fourth six counties with their chief towns?

A. Kincardine,	Stonehaven.
Aberdeen,	Aberdeen.
Inverness,	Inverness.
Nairn,	Nairn.
Cromarty,	Cromarty.
Fife,	Cupar.

Q. What are the names of the remaining nine counties with their chief towns?

Counties	Chief Towns.
A. Forfar,	Forfar.
Banff,	Banff.
Sutherland,	Dornoch.
Clackmannan,	Clackmannan.
Kinross,	Kinross.
Ross,	Tain.
Elgin,	Elgin.
Orkney,	Kirkwall.
Berwick,	Dunse.

Q. Where and what is the town of Berwick ?

A. The town of Berwick stands on the borders of England and Scotland, properly belonging to neither ; it is very large, populous, and well built.

Q. What is the population of Scotland?

A. In 1841 the population amounted to two millions six hundred and twenty thousand.

Q. What are the chief cities of Scotland ?

A. Edinburgh, (the capital), Glasgow, Aberdeen, and Perth.

Q. What are the islands of Scotland.?

A. The Islands belonging to Scotland are the Shetland, the Orkney, Hebrides, Skye, Arran and Bute, &c.

Q. What are the principal lakes in Scotland ?

A. The chief lakes of Scotland are 'Loch Tay, Loch Lomond, Loch Ness, and Loch Awe.

Q. What are its highest mountains?

A. The highest mountains of Scotland are Ben Nevis, Ben Lomond and Ben Macdhui.

CHAPTER XXII.

Of Ireland.

Q. How is Ireland divided ?

A. Into four provinces, namely, Ulster northward ; Leinster eastward ; Munster southward ; and Connaught westward ; which provinces are divided into thirty-two counties.

Q. What are the names of the eight first with their chief towns ?

Counties.	Chief Towns.
A. Dublin,	Dublin,
Louth.	Drogheda.
Wicklow,	Wicklow.
Wexford,	Wexford.
Longford,	Longford.
East Meath,	Trim.
West Meath,	Mullingar.
King's County,	Philipstown.

Q. What are the names of the second eight counties with their chief towns?

A. Queen's County,	Maryborough.
Kilkenny,	Kilkenny.
Kildare,	Kildare.
Carlow,	Carlow.
Down,	Down.
Armagh,	Armagh.
Monaghan,	Monaghan.
Cavan,	Cavan.

Q. What are the names of the third eight counties with their chief towns?

A. Antrim,	Carrickfergus.
Londonderry,	Derry.
Tyrone,	Omagh.
Fermanagh,	Enniskillen.
Donegal,	Lifford.
Leitrim,	Leitrim.
Roscommon,	Roscommon.
Mayo,	Castlebar.

Q. What are the names of the last eight counties with their chief towns?

A. Sligo,	Sligo.
Clare,	Ennis.
Cork,	Cork.
Kerry,	Tralee.
Limerick,	Limerick.
Tipperary,	Clonmel.
Waterford,	Waterford.
Galway,	Galway.

Q. What is the capital of Ireland?
A. Dublin.
Q. What are the principal rivers in Ireland?
A. The Shannon, the Blackwater, the Liffy, and the Boyne.

Q. How many inhabitants does Ireland contain?
A. About eight millions.
Q. How many members are sent to the British Parliament by England and Wales?
A. Five hundred members are sent to the British Parliament.
Q. How many are sent by Scotland?
A. Scotland sends fifty-three members to the British Parliament.
Q. How many are sent by Ireland?
A. Ireland sends one hundred and five members to the British Parliament, making in all six hundred and fifty-eight members.

CHAPTER XXIII.

Of Asia.

Q. What are the principal divisions of Asia?
A. Asia, one of the largest and richest quarters of the globe, contains Asiatic Turkey, Siberia, China, Tonquin, the Birman Empire, Hindostan, Persia, Independant Tartary, Arabia, and Japan.
Q. Describe the climate and soil of Asia?
A. In a country of such immense extent as Asia both the soil and climate must be extremely various, but in general it is far superior to Europe and Africa both in the serenity of its air, and the fertility of its soil, producing the most delicious fruits, the most fragrant and balsamic plants, spices, and gums.
Q. In what else is Asia superior to Europe and Africa?
A. In the quantity, variety, beauty, and value of its gems, the riches of its metals, and the fineness of its silks and cottons.
Q. For what is Asia peculiarly celebrated?
A. 1st. Here the first man was created, the Patriarchs lived, the law was given to Moses, and here the great and celebrated empires of Assyria, Babylon, and Persia were also formed.

B 2

2ud. In Asia, Christianity was first promulgated, and thence the light of the Gospel was diffused over the whole world.

3rd. The laws, arts, and sciences almost all had their origin in Asia. At Ephesus, in Natolia, stood the celebrated temple of Diana, burnt the night on which Alexander the Great was born. In Asia was the famous city of Troy. Tyre and Sidon were on the coast of Palestine, formerly great cities of trade, but now inhabited by a few fishermen. Sodom and Gomorrah, places once famous, are now only known by name; and Nineveh and Babylon, cities in Mesopotamia near Bagdat, are now the habitations of wild beasts.

Q. What are the principal mountains of Asia?

A. The Asiatic mountains are Ararat near the Caspian Sea; Lebanon in Judea; Taurus, and the lofty Caucasus.

Q. What are its chief rivers?

A. The Euphrates, the Tigris, the Indus, and the Ganges.

Q. What are the principal islands of Asia?

A. The chief Asiatic islands are the Japanese isles, the Ladrones, the Philippines, the Molucca or Clove isles, the Banda or Nutmeg isles; Borneo, Sumatra, and Java; Ceylon, Nicobar isles; New Holland, Van Dieman's Land, New Guinea, the Pelew isles, and the Carolines, with some others of less note.

CHAPTER XXIV.

Of Asiatic Turkey.

Q. For what is Asiatic Turkey particularly celebrated?

A. As the principal scene of early Scripture history; and of those romantic expeditions called the Crusades.

Q. What is a Crusade?

A. A holy war; a name given to the expeditions

of the Christians against the infidels for the recovery of Palestine, so called because those who were so engaged in them, wore a *Cross* on their clothes, and bore one on their standard.

Q. What else have you to observe of this country ?

A. That it is covered with wrecks of ancient splendour ; it was once very fertile and rich, containing many flourishing kingdoms, and a great number of celebrated cities ; but it is now quite barren, the people miserable, and the cities few.

Q. How is Asiatic Turkey divided ?

A. Into nine provinces, namely, Asia Minor, Caramania, Roum, Armenia, Kurdistan, Irak-Arabia or Chaldea, Algesira or Mesopotamia, Syria, and the Holy Land.

Q. What are the principal cities in Turkey in Asia ?

A. The principal cities of Turkey in Asia, are Smyrna, Aleppo, Damascus, Jerusalem, and Bagdat. Damascus is a very extensive and populous city, containing about 140,000 inhabitants.

Q. What are its chief mountains ?

A. the chief mountains of Turkey in Asia are Taurus, Ararat, and Lebanon.

Q. For what is Turkey in Asia particularly celebrated ?

A. Turkey in Asia is celebrated for its carpets, oils, silks, rhubarb, and fruits.

CHAPTER XXV.

Of Asiatic Russia.

Q. What kind of country is Russia in Asia ?

A. It is a very large country, comprehending the most northern parts of Asia; being in length, from the Black Sea to the southern extremity of Kamtschatka, 4,480 miles, and in breadth from north to south 1,800 miles.

Q. How is this country divided ?

A. Into five governments, namely, Western Siberia; Eastern Siberia, Orenburg, Astracan, and Caucasus.

Q. What is the population of Asiatic Russia ?

A. It is estimated at eight millions.

Q. What is the produce of Asiatic Russia ?

A. The southern part produces all the necessaries of life, but the northern part is extremely cold, almost uncultivated, and thinly peopled. The principal riches of the country consist of fine skins and furs.

Q. Which are the principal cities in Asiatic Russia.

A. Astracan, Tobolsk, and Orenburg.

Q. How far is Astracan from London ?

A. 2,149 miles.

Q. What are the names of the principal nations of Asiatic Russia ?

A. The Cossacks, Calmucks, Circassians, and Georgians, the Samoieds, Ostiaks, Koriaks, the Tschutki, Kamtschatdales and others of less note.

Q. Are the Asiatic Russians like those of Europe ?

A. No, very different ; for in this wide expanse there are many distinct races of men, not only differing from the European Russians, but also from each other, both in appearance, manners, religion, and language.

Q. What nations live towards the north ?

A. Samoieds, Ostiaks, Koriaks, and other similar tribes, who are sunk in gross superstition and idolatry, and very filthy and squalid in their appearance.

Q. How do they subsist ?

A. Chiefly by hunting and fishing, but are all averse to steady industry. Some are fixed, and others wandering. These people in general are of small stature and hardfeatured ; and considering the extent of country they occupy, are few in population.

Q. Who are the *Tschutki* * ?

A. The Tschutki, who inhabit the north-east corner of Russia, are superior in size, and better featured than the preceding tribes, and are more skill-

* Pronounced Chutki.

ed in the various arts of life. The peninsula of
Kamtschatka is inhabited by another race, but not
less filthy and brutalized than the Samoieds. Several of the tribes before mentioned are considered in general very dull, heavy and harmless.

Q. Which are the chief nations inhabiting the
southern parts of Asiatic Russia ?

A. The Cossacks, Circassians, and Georgians.

Q. Describe the Cossacks ?

A. The Cossacks are divided into the Don Ukrain
and Uralian Cossacks ; but the most celebrated are
those who reside on the borders of the river Don.
It was the latter who under their enterprising leader,
Platoff, made such dreadful havoc among the French
in Bonaparte's unsuccessful campaign into Russia
in 1812, when he lost nearly 300,000 men.

Q. What are the Calmucks ?

A. The Calmucks are a nation of wanderers, who
live in tents, and remove thence in quest of pastur-
age for their numerous cattle, consisting of horses,
camels, cows, and sheep. These people neither sow
nor reap, so that they live without bread or any
kind of vegetable. Their food is fish, flesh, milk,
butter and cheese.

Q. Who are the Circassians ?

A. The Circassians are a noble people, very courageous, and possessing a great military genius, and
were they united under one chief, might become a
great independent empire ; but they are a nation of
wandering mountaineers, divided into many different and hostile tribes, and want a spirit of unity to
make their power effectual.

Q. What is Georgia ?

A. Georgia is a very fine country, situated between the Black and Caspian Seas· It is very productive, and the inhabitants are a fine and warlike
people ; its chief town is Tiflis.

CHAPTER XXVI.

Of China.

Q. What is China ?

A. China is one of the most ancient and polished nations of Asia, and is celebrated for its extent and prodigious population.

Q. For what is it otherwise celebrated?

A. For its immense wall, which is one thousand two hundred miles long, twenty-five feet high, and eighteen thick.

Q. For what purpose was this immense wall built?

A. To divide China from Tartary, and keep the Tartars from plundering the Chinese territories.

Q. What are the chief cities of China?

A. Pekin, the capital; Nankin; and Canton.

Q. What is the character of the Chinese?

A. They are industrious beyond any people on earth; possess a great share of ingenuity, and are generally honest in their dealings. They are lovers of the arts and sciences, but have too great an opinion of their own wisdom, and think meanly of other nations.

Q. What else is worthy of observation in respect to the Chinese?

A. Their complexion, which is tawny; and those are thought to excel in beauty who are most bulky. The women affect much modesty, and are remarkable for their small feet.

Q. How far is Pekin from London?

A. 5,000 miles.

Q. What is the population of China?

A. One hundred and fifty millions.

Q. What is Thibet?

A. Thibet, sometimes called the Snowy Region, is a vast extent of country in the interior of Asia, subject to China?

Q. How is it divided, and what is the number of its inhabitants?

A. Thibet is divided into three parts, Upper, Middle, and Lower; its inhabitants are estimated at five millions.

Q. What is Chinese Tartary?

A. Chinese Tartary is a large country to the north

of China Proper, to which it is tributary, whose population consists of about six millions. The several
countries, namely, China Proper, Chinese Tartary,
and Thibet, constitute what is called the " *Chinese
Empire.*"

CHAPTER XXVII.
Of India in General.

Q. What is meant by India?

A. The general name of India is now applied to
those vast regions of Southern Asia on the confines
of China, and from the mountains of Thibet in the
north to the Ocean.

Q. Why is it called India?

A. It is so called from the river Indus.

Q. How is it divided?

A. India is generally divided into two great
parts, namely, India Interior and India Exterior; or
India within and India without the Ganges.

Q. What kind of country is India?

A. India consists chiefly of extensive plains, ferti-
lized by a great number of beautiful meandering
rivers and purling streams, and interspersed with a
few ranges of hills. The periodical rains and in-
tense heats produce a luxuriance of vegetation almost unknown to any other country in the world.

Q. What is the soil of India?

A. It is rich in every kind of production, whether
fossil, vegetable, or animal. The Indians sow the
ground in May and June before the rainy season,
and reap in November and December, which are the
most temperate months in the year.

Q. What are the productions of India?

A. Wheat, rice, barley, and other grain, in great
plenty and perfection ; also, all kinds of fruits, spices, &c.

Q. What is the character of the Hindoos?

A. The Hindoos are extremely mild, and by no
means adapted for hard labour or war.

Q. What is the population of Hindostan ?
A. One hundred and thirty-four millions.
Q. What is the capital of British India ?
A. Calcutta.
Q. How far is Calcutta from London ?
A. 4.929 miles.
Q. What other remarkable cities are there?
A. Delhi, Madras, Pondicherra, Seringapatam, Goa, and Bombay.
Q. What are the chief mountains ?
A. The Gauts, or Ghats ; and the Himmalaya in the north, the highest in the world.
Q. What are the chief mines of India?
A. The chief mines of India are gold, silver and diamonds.
Q. What are the chief animals of India ?
A. They are elephants, rhinoceroses, tigers, leopards, panthers, camels, dromedaries, buffaloes, and monkeys.

CHAPTER XXVIII.

Of Interior India, or Hindostan.

Q. How is Hindostan situated ?
A. Hindostan is bounded N. by Thibet, E. by part of the Birman Empire, and the Bay of Bengal, S. by the Indian Ocean, W. by Persia and the Arabian Sea.
Q. How are the inhabitants divided ?
A. The Hindoos are divided into tribes or castes. The principal are the Brahmins, soldiers, mechanics, and labourers. The Brahmins have the care of religion; the soldiers are usually called rajahpoots ; those in the service of the English are called Sepoys ; the mechanics comprehend merchants and traders ; the labourers include farmers and all who cultivate the land. The greater part of the Hindoos are idolators.

Of Exterior India.

Q. What is Exterior India ?

A. Exterior India is a vast extent of country, comprising the Birman Empire, Malacca, Siam, and the Empire of Tonquin.

Q. How are these countries situated ?

A. The Birman Empire, Malacca, Siam, and Tonquin, are situated to the east of 'the Bay of Bengal, and are bounded on the south and east by the Ocean, and on the north by the Chinese Empire.

, CHAPTER XXIX.

Of the Birman Empire.

Q. What is the situation of the Birman Empire ?

A. The Birman Empire is situated east of Hindostan, south of China, and west of the Chinese Sea.

Q. What are its principal divisions ?

A. Its principal divisions are Ava, Pegu, Laos, Cambodia, and part of Siam.

Q. What is the general character of the Birmans ?

A. They are a lively and inquisitive race of people, but passionate and revengeful.

Q. What are its chief cities ?

A. The chief cities of the Birman Empire are Ava, its ancient, and Ummerapoora, its present capital.

Q. How far is Ava from London?

A. 5,346 miles.

Q. What is the population of the Birman Empire ?

A. Seventeen millions.

Q. What are the chief productions of Birmah ?

A. The soil being extremely fertile, it yields excellent wheat, sugar, canes, indigo, and cotton ; and the different tropical fruits in high perfection, and in great abundance. Its animals are the same as those of Hindostan. It possesses excellent mines of gold, silver, and precious stones.

CHAPTER XXX.

Of Siam.

Q. What kind of country is Siam?

A. Before the recent extension of the Birman Empire, this country was considered the chief nation of Exterior India ; but, a great part of these dominions having been conquered by the former power, it is now but of small extent, being about 700 miles in length, and 200 in breadth, and containing about eight millions of inhabitants.

Q. What is its chief city?

A. Its chief city is Siam, beautifully situated, about fifty miles from the sea, on the river Meinam, containing about 100,000 inhabitants.

Q. In what do the Siamese excel?

A. The Siamese excel in some of the ornamental manufactures, particularly in those of gold and silver, and their miniature paintings have been much admired.

Q. Describe the Siamese?

A. The Siamese in general are of a dark complexion ; and their food consists principally of rice and fish. The men are extremely indolent, and most of the laborious works is executed by the women.

CHAPTER XXXI.

Of Malacca.

Q. What is Malacca?

A. Malacca, a large peninsula of Asia, is divided into two kingdoms; that of Patani in the north, and Yahor in the south.

Q. What are the chief cities of Malacca?

A. The chief city of Yahor is Malacca, and that of Patani, Queda. The former contains about 12,000, inhabitants.

Q. Describe the inhabitants?

A. The Malays are in general of small stature, but well made, with a tawny complexion, and long black shining hair. These people are classed among the principal merchants of the East, and their colonies and commerce are widely spread all over the Indian Seas.

Q. What is their general character?

A. The Malays are deeme 1 the most treacherous and most ferocious people on the globe, and their history is full of such enterprises as mark this desperate character.

CHAPTER XXXII.
The Empire of Tonquin.

Q. Describe the Empire of Tonquin.

A. The name of this empire is but of recent date; it was formerly tributary to China. It now comprises Tonquin, Cochin China, Siam, and Laos. For its size, it is one of the finest kingdoms of the east, having a population of about 21,000,000 of inhabitants; and possessing a fruitful soil, and a mild and salubrious climate.

Q. What are its chief productions?

A. They are very numerous, and the fruits excel all the rest of Asia in flavour and beauty; and their elephants are esteemed the largest and swiftest in the world.

Q. What is its chief city?

A. The chief city is Backu, formerly called Kesho, containing 40,000 inhabitants.

CHAPTER XXXIII.
Of Persia.

Q. What kind of country is Persia?

A. Persia is a large country, containing about nine millions of inhabitants Its chief cities are Ispahan and Teheran.

Q. How is Persia situated?

A. It is bounded N. by the Caspian Sea and part

of Tartary ; E. by India ; W. by Asiatic Turkey ; and S. by the Ocean and Persian Gulf.

Q. What are its climate and soil ?

A. It is said that no country in the world has greater variety of climate and soil than this, being extremely mountainous, barren, and cold in the northern parts ; in the middle, very mild and sandy, but enjoying a temperate air ; and in the south, level and extremely fertile, though for several months the heat is excessive.

Q. What is the character of the Persians?

A. They are very luxurious and effeminate, pos· sess much good sense and ability, are honest in their dealings, but passionate and revengeful.

Q. What are the chief productions of Persia ?

A. Corn, wine, oil, great abundance of oranges, dates, melons, grapes, and nuts ; also, senna, rhubarb, and various other drugs ; and it is famous for silks.

Q. How do the Persians write?

A. The Persians write from the right hand to the left, and, as no printing is allowed, a great number of people are constantly employed in writing, an art in which they excel.

Q. What is their religion ?

A. The greater part of the Persians are Mahomedans, with some idolaters, who worship the sun, fire, &c.

Q. What are the chief towns ?

A. Casbin, Tauris, and Shiraz.

Q. For what are these towns famous?

A. *Casbin* is large and populous, and is noted as the mart for almonds, raisins, and melons ; *Tauris*, the second city in Persia, is noted for its mosques and caravansaries, and is also famous for its great trade in cotton, cloths. silks, and gold and silver brocades ; the population is upwards of 550,000. *Shiraz* is also very large and populous, and is famous for its wines.

CHAPTER XXXIV.

Of Independent Tartary.

Q. What kind of country is Tartary?

A. Tartary is of very great extent, situated on the north of India.

Q. What is the character of th' Tartars?

A. The generality of the Tartars are wanderers like their ancestors, the Scythians, and are celebrated as a hardy race, who live in tents and lead a roving life, and with their families, their cattle, and all they possess, wander about in the temperate parts and vast solitudes of the Asiatic Continent from the confines of Europe to the great Pacific Ocean. On meeting with a fertile and pleasant situation, they pitch their tents, and continue in the place till the grass is all consumed, when they again set forward in search of another inviting spot.

Q. What is the chief city of Tartary?

A. Samarcand.

Q. How far is Samarcand from London?

A. 3,127 miles.

Q. What is supposed to be the population of Tartary?

A. Ten millions.

CHAPTER XXXV.

Of Arabia.

Q. What kind of country is Arabia?

A. Arabia is a large country, situated East of the Red Sea, and South of Turkey. It contains about ten millions of inhabitants.

Q. What is the general character of the Arabians?

A. The Arabs generally live under tents; the greater part are wandering tribes, like the Tartars, committing depredations and robberies wherever they go.

Q. What are the chief productions of Arabia?

A. The most valuable productions are its horses, camels, gum, and coffee. The horses are said to be the best in the world.

Q. What are its chief towns ?

A. Mecca, where Mahomet, the founder of the Mahomedan religion, was born ; and Medina, where he was buried.

Q. How far is Mecca from London ?

A. 3,988 miles.

CHAPTER XXXVI.

Of Japan.

Q. What kind of country is Japan ?

A. Japan is a very large country, situated east of Asia with the title of an empire, composed of a great number of islands difficult of access, the principal of which is Niphon.

Q. For what is Japan celebrated ?

A. Japan is said to be the richest country in the world for gold ; and the air and water are most excellent.

Q. What is the general character of its inhabitants?

A. They are naturally ingenious, and possess a good memory, but their manners are exactly opposite to ours. Our common drinks are cold, theirs are hot ; we uncover the head out of respect, and they the feet ; we are fond of white teeth, and they of black : we get on horseback on the left side, and they on the right ; and they have a language so peculiar that it is not to be understood by another nation.

Q. What is the population of Japan ?

A. Thirty millions.

Q. What is its chief city ?

A. Jeddo.

Q. How far is Jeddo from London ?

A. 5,942 miles.

CHAPTER XXXVII.

Of Africa.

Q. What is Africa?

A. One of the four quarters* of the world, bounded N. by the Mediterranean Sea, W. and S. by the Ocean ; and E. by the Red Sea and the Isthmus of Suez.

Q. What are its climate and soil?

A. The greatest part of Africa, lying under the Torrid Zone, the heat is almost insupportable in many places. However, the coasts in general are very fruitful, the fruits excellent, and the plants extraordinary.

Q. Are there not several deserts in Africa.

A. Yes; several of which are almost without water, and whose sands are so loose that, by means of a strong wind, they will sometimes bury whole caravans at a time.

Q. What are its productions?

A. Its productions are gold, fruit, gum, &c., camels, elephants, and all sorts of wild beasts.

Q. What are the principal rivers?

A. The two largest are the Nile and the Niger, which annually overflow their banks and fertilize the adjacent countries.

Q. How is Africa divided?

A. Africa is divided into the following general parts, viz.: Barbary, Egypt, Zaara or the Desert, Negroland, Guinea, and Ethiopia.

CHAPTER XXXVIII.

Of Barbary.

Q. How is Barbary divided?

A. Barbary is divided into five states, namely, Morocco, Algiers, Tunis, Tripoli, and Barca.

* By modern geographers, the world is now divided into five great portions, instead of four quarters as formerly. (See p. 4.)

Q. What are the situations and boundaries of M(·rocco?

A. Morocco is about 500 miles in length and 48 in breadth. It is bounded N. by the Mediterranf; Sea; E. by Algiers; S. by the desert of Tafilet; ar. W. by the Atlantic Ocean. Morocco is its capital.

Q. What is the situation of Algiers?

A. Algiers is about 480 miles in length, and 70 ; i breadth. It is bounded N. by the Mediterranean. S. by Mount Atlas: E. by Tunis; and W. by the Empire of Morocco. Algiers is its capital.

Q. What is the situation of Tunis?

A. Tunis is 202 miles in length, and 170 in breadth. It is bounded N. by the Mediterranean; S. by Tripoli; and W. by Algiers. Tunis, its capital, is built near the ruins of the once famous Carthage.

Q. What is the situation of Tripoli?

A. Tripoli is situated on the borders of the Mediterranean, between Barca on the east, and Tunis on the west.

Q. What is the situation of Barca.

A. Barca is situated between Tripoli and Egypt. Its chief city is Barca.

CHAPTER XXXIX.

Egypt.

Q. How is Egypt situated?

A. Egypt is bounded N. by the Mediterranean; E. by the Red Sea; S. by Nubia and Abyssinia; and W. by Barca.

Q. How is Egypt divided?

A. Egypt is divided into three great parts called the Upper, Middle, and Lower. Upper Egypt is the southern part, and Lower Egypt the northern.

Q. Which are the chief towns in Egypt?.

A. Grand Cairo and Alexandria.

Q. How far is Grand Cairo from London ?

A. 2,188 miles.

Q. Describe the soil and climate of Egypt ?

A. The climate of Egypt is naturally hot and unwholesome ; but the soil is exceedingly fruitful, owing to the overflowing of the Nile, which leaves a fattening manure behind.

Q. For what is Egypt remarkable ?

A. For its pyramids, which are stupendous fabrics that have stood several thousand years.

Q. To whom does Egypt belong ?

A. Egypt is a part of the Turkish Empire, and the Sultan is its nominal Sovereign, and receives tribute : but by treaty Mehemet Ali and his family are invested with it as a Pachalic, with authority nearly amounting to independence.

CHAPTER XL.

Of Zaara, or the Desert.

Q. How is Zaara bounded ?

A. Zaara is bounded N. by the Barbary States, E. by Egypt and Nubia, S. by Negroland and Guinea, and W. by the Atlantic Ocean.

Q. Who are the inhabitants ?

A. Zaara is inhabited by tribes of rambling Arabs, who have but few towns or settled dwellings that deserve description.

Of Negroland, Nigritia, or the Country of the Blacks.

Q. What are the boundaries of Negroland ?

A. It is bounded N. by Zaara ; E. by Nubia ; S. by Guinea ; and W. by the Atlantic.

Q. How is it divided ?

A. Negroland is divided into a great many different nations, whose limits are but little known.

Q. What is the character of the Negroes ?

A. The Negroes are mostly an uncivilized and ignorant people, and are said to be very crafty towards the Europeans.

C

CHAPTER XLI.

Of Upper Guinea

Q. How is Upper Guinea bounded ?
A. Upper Guinea is bounded N. by Negroland, E. by unknown parts, S. and W. by the Ocean.
Q. How is it divided ?
A. Into the Grain Coast, the Ivory Coast, the Gold Coast and the kingdoms of Benin, Ashantee, &c.
Q. What is the character of the natives ?
A. They are generally courteous to strangers, and do not want good sense ; but they are said to be addicted to cheating.

CHAPTER XLII.

Of Lower Guinea.

Q. What countries does Lower Guinea comprehend ?
A. It comprehends five principal kingdoms, namely, Congo, Loango, Angola, Matamba, and Benguela.
Q. What are the capitals of each ?
A. St. Salvador is the capital of Congo ; Loango, of Loango ; Loanda, of Angola ; Santa Maria, of Matamba ; and Benguela, of Benguela.
Q. How are these kingdoms bounded ?
A. They are bounded W. by the Atlantic, but on the N. E. and S. by countries unknown to Europeans.
Q. By whom is the trade principally possessed ?
A. The trade of Lower Guinea is chiefly engrossed by the Portuguese, whose principal fort is Loanda, called by them St. Paul, a very neat and handsome city, containing 3000 houses built of stone, besides a great number of habitations of a meaner kind possessed by the Negroes.

CHAPTER XLIII.
Of Ethiopia.

Q. What is Ethiopia?
A. Under the general name of Ethiopia are included all the remaining parts of Africa excepting those which are unexplored.
Q. How is it divided?
A. Into two great parts, Upper and Lower.
Q. What countries does Upper Ethiopia comprehend?
A. Upper Ethiopia comprehends Nubia, Abyssinia, and the coasts of Abex and Zanguebar.
Q. What countries does the Lower contain?
A. Lower Ethiopia comprehends Caffraria, Monomotapa, Monumugi, and Sofala, the country of the Hottentots, and the colony of the Cape of Good Hope.
Q. What is the capital of Nubia?
A. Sennaar.
Q. What is the capital of Abyssinia?
A. Gondar.
Q. How far is Gondar from London?
A. 3,370 miles.
Q. What is the population of Abyssinia?
A. Two millions.
Q. What is the population of the empire of Morocco, including Fez, Tafilet and Susa?
A. Nearly fifteen millions.
Q. What is the population of Algiers?
A. A quarter of a million.
Q. What is the population of Tunis?
A. A quarter of a million.
Q. What is the population of Tripoli?
A. Half a million.

CHAPTER XLIV.
Of African Islands.

Q. What are the principal islands belonging to Africa?
A. The Canaries, Madagascar, Mauritius or Isle of

France, Bourbon, and St. Helena, Madeira, Cape de Verd Isles, and the Azores.

Q. Describe the Canaries ?

A. The Canaries, seven in number, are very fruitful in corn, wine and fruits. They belong to Spain, and are celebrated as having produced the canary-bird.

Q. Describe the Azores and Madeira ?

A. The Azores, nine in number, are very fertile and salubrious. Madeira is renowned for its excellent wine. These islands belong to Portugal.

Q. Describe St. Helena ?

A. This small but important island is possessed by the English. It contains about three thousand of inhabitants, and is remarkable for the exile and death of Bonaparte.

Q. What is Madagascar ?

A. Madagascar, in the Indian Ocean, is one of the largest islands in the world, and is considered very fruitful. It is supposed to contain many independent states, but the interior is very little known.

Q. Describe the isles of Bourbon and France ?

A. These islands are in high cultivation, rich and fruitful, producing all the necessaries of life. The Isle of France contains about seventy thousand inhabitants, and Bourbon, sixty-eight thousand. The former belongs to the English, and the latter to the French.

CHAPTER XLV.

AMERICA.

Q. What is America ?

A. One of the five great portions of the world and by much the largest. It is bounded on all sides by the Ocean, as appears from the latest discoveries.

Q. How is it divided ?

A. Into North and South America.

Q. What are the principal states of North and South America?

A. North America comprehends the United States, British America, and Mexico. South America, Brazil, Paraguay, Chili, Peru, Amazonia, Patagonia, and Guiana.

Q. What are its chief mountains?

A. The Stony and Alleghany in North, and the Andes or Cordilleras in South America.

Q. What are its chief lakes?

A. Lakes Champlain, Superior, Huron, Michigan, Erie, and Ontario.

Q. What are the chief bays?

A. Baffin's and Hudson's Bay, and the Gulf of Mexico.

Q. What are its principal rivers?

A. Those of the St. Lawrence, the Hudson, the Mississippi, and the Missouri, in North America; and the rivers of the Amazons, and the Rio de la Plata, in South America.

Q. Who discovered America?

A. Columbus, a Genoese, in 1492.

CHAPTER XLVI.

Of the United States.

Q. What part of North America is possessed by the United States?

A. All the eastern parts, extending westward to the Pacific.

Q. How are they divided?

A. Into four parts; Northern, Middle, Southern, and Western.

Q. What are these States?

A. The Northern States are Maine, New Hampshire, Vermont, Massachusetts, Rhode Island, and Connecticut. The Middle States are New York, Pennsylvania, New Jersey, and Delaware. The Southern States are Maryland, Virginia, North Carolina, South Carolina, Georgia, Florida, Alabama,

Mississippi, Louisiana, and Texas. The Western States are Arkansas, Tennesse, Kentucky, Ohio, Michigan, Indiana, Illinois, Missouri, Iowa, Wisconsin, and California. Besides these there is the District of Columbia, in which the city of Washington, the capital of the United States, is situated; and the following territories, not yet erected into States, namely, Oregon, Minesota, Utah, and New Mexico.

Q. What are the productions of the United States?

A. They produce abundance of cotton, grain, fruit, tobacco, leather, skins cattle, timber, hemp, flax, and all kinds of metals.

Q. What is the general character of the inhabitants of the United States?

A. They are reputed to be proud and independent. Frugality, industry and attachment to liberty, are the leading parts of their character.

Q. What is the capital of the United States?

A. Washington, 3,658 miles from London.

Q. What are the chief ports?

A. Boston, Philadelphia, New York, and Charleston.

Q. What is the population of the United States?

A. Its inhabitants amount to about seventeen millions, of whom two and a half millions are slaves.

CHAPTER XLVII.

Of Mexico.

Q. What were the Spanish dominions in North America?

A. California and Mexico.

Q. What political changes took place in these countries?

A. They threw off the yoke of Spain, and formed the independent republic of Mexico.

Q. What are the chief ports of Mexico?

A. Acapulco on the Pacific, and Vera Cruz on the Gulf of Mexico.

Q. What are the soil and climate ?

A. The climate in general is pure and salubrious, and the soil is extremely fertile. Mexico is it capital.

CHAPTER XLVIII.

Of the British Possessions in North America.

Q. What are the possessions in North America which belong to Great Britain ?

A. The British Dominions are very extensive, including the Hudson's Bay Territories, Canada, New Brunswick, Nova Scotia, the Islands of Newfoundland, Cape Breton, Prince Edward, and the Bermudas, besides several smaller islands in the Gulf of St. Lawrence.

Q. Describe the climate and soil ?

A. The climate is very severe for a great part of the year, but it is nevertheless healthy. The soil is equal to any in the world, and under proper cultivation yields large crops of all kinds of grain, hay, potatoes, &c. The trade in timber, furs, pot and pearl ashes, and the fisheries, makes these colonies very valuable.

Q. What is the population of British America?

A. It is estimated to amount to 2,472,195.

Q. What are the principal rivers and lakes in British America?

A. The principal rivers are the St. Lawrence, Ottawa, St. Francis, St. Maurice, St. John, Niagara, Coppermine, and Mackenzie. The principal lakes are Ontario, Huron, Erie, Superior, Athabasca, and Great Slave Lake.

Q. What is the length of British America ?

A. From Cape Charles to the North Pacific it is 3,500 miles long.

Q. What is its breadth ?

A. From Barrow's Strait to the Missouri territory it is 2,000 miles in breadth.

* The Re-Union of Upper and Lower Canada was proclaimed by the Governor on the 10th February, 1841.

Q. What are the names of the chief cities and towns in Canada?

A. Quebec (the Seat of Government), Montreal, Three Rivers, Sherbrooke, Stanstead, St. John's, Chambly, Sorel, St. Eustache, and L'Assomption, in Eastern Canada ; and Toronto, Kingston, Cornwall, Brockville, Prescott, Bytown, Cobourg, Hamilton, Niagara, London, Sandwich, and Brantford, in Western Canada.

Q. How is Eastern (formerly Lower) Canada divided?

A. Into the three Principal Districts of Quebec, Montreal, and three Rivers, and the two inferior Districts of Gaspé, and St. Francis, all of which are subdivided into the 36 Counties* of—1. Vaudreuil, 2. Ottawa, 3. Lake of Two Mountains, 4. Montreal, 5. Terrebonne, 6. Leinster, 7. Berthier, 8. St. Maurice, 9. Champlain, 10. Port Neuf, 11. Quebec, 12. Montmorency, 13. Saguenay, 14. Gaspé, 15. Bonaventure, 16. Rimouski, 17. Kamouraska, 18. L'Islet, 19. Bellechasse, 20. Dorchester, 21. Megantic, 22. Lotbiniere, 23. Nicolet, 24. Yamaska, 25. Drummond, 26. Sherbrooke, 27. Stanstead, 28. Missisquoi, 29. Shefford, 30. St. Hyacinthe, 31. Rouville, 32. Richelieu, 33. Vercheres, 34 Chambly, 35. Huntingdon, 36. Beauharnois.

Q. What share has Eastern Canada of the Representation in the United Provincial Parliament?

A. Each County returns one Member, who, added to two each from the cities of Quebec and Montreal, and one each from the Towns of Three Rivers and Sherbrooke, make the share of Lower or Eastern Canada in the representation of the Provincial Legislature to amount to 65.

Q. How is Western (formerly Upper) Canada divided?

A. Into the following Districts and Counties, viz:—

* The Counties are divided into Seigniories and Townships. The Parishes sometimes contain but one Seigniory ; sometimes, on the contrary, a Parish is formed of several Seigniories or Townships, either entire or divided.

Districts,	Counties.
Bathurst,	Lanark and Renfrew.
Brock,	Oxford.
Colborne,	Peterborough.
Dalhousie,	Carleton.
Eastern,	Stormont, Dundas, and Glengarry
Gore,	Wentworth (including the Townships of Seneca and Oneida) and the County of Halton.
Home,	York and the City of Toronto.
Huron,	Huron.
Johnstown,	Leeds and Grenville.
London,	Middlesex.
Midland,	Frontenac, Lennox, and Addington.
Newcastle,	Northumberland and Durham.
Niagara,	Lincoln and Welland, and the County of Haldimand, except the Townships of Seneca, Oneida, Rainham, and Walpole.
Ottawa,	Prescott and Russell.
Prince Edward,	Prince Edward.
Simcoe,	Simcoe.
Talbot.	Norfolk.
Victoria,	Hastings.
Wellington,	Waterloo.
Western	Essex and Kent.

Q. What share has Western Canada of the Representation in the United Provincial Parliament?

A. The Counties and Ridings return each one Member to Parliament, who with two from the City of Toronto and one each from Kingston, Cornwall, Brockville, Hamilton, Niagara, Ottawa, and London, make the share of Upper or Western Canada in the representation of the Legislature to be 65.

Q. What are the names of the chief cities and towns of other British Provinces in North America?

A. Fredericton and St. John are the principal towns of New Brunswick; Halifax, of Nova Scotia; St. John of Newfoundland; Sydney, of Cape Breton; Charlotte town, of Prince Edward Island; and St. George, of the Bermudas.

Q. How are the British American Provinces governed?

A. They have each a governor, who is appointed by the Sovereign of England, and represents Her

Majesty in the Colony to which he is sent ; a Legislative Council, appointed by the Sovereign ; and a
House of Assembly, elected by the people. It is
necessary, before any laws can be binding, that they
be passed by the Legislative Council and the House
of Assembly, and receive the assent of the Governor
in the name of the Sovereign.

Q. When was Canada first settled, and by what
people ?

A. It was first settled in 1608 by the French, and
they continued to possess it until 1759, when Great
Britain acquired it by conquest, and has ever since
possessed it.

CHAPTER XLIX.

Of the former Spanish South America.

Q. What parts of South America once belonged to
Spain ?

A. Terra Firma, Peru, Chili, and Paraguay, which
produce gold, silver, jewels, gums, and drugs, and
the choicest fruits.

Q. What political changes have occurred, and
what are the present divisions ?

A. Like Mexico, these provinces have asserted
their independence ; and are now divided into the
republics of Columbia, Peru, Chili, and the United
Provinces of La Plata.

Q. What is Columbia ?

A. It contains the N. part of South America. The
capital is Bogota; and the chief ports, Carthagena,
Caraccas, and Porto Bello.

Q. What is the capital of Peru ?

A. Lima.

Q. How far is Lima from London ?

A. 5,700 miles.

Q. What are the chief ports of Peru ?

A. Callao near Lima, Guayaquil, and Truxillo.

Q. In what does Peru chiefly abound ?

A. Gold and silver mines.

Q. What is Chili, and what is its capital ?

A. Chili is to the S. of Peru. Its capital is St. Jago, and its chief ports are Valparaiso and Conception.

Q. What are the United Provinces of La Plata?

A. They are situated between Peru and Chili, and the Atlantic, and consist of Paraguay, Buenos Ayres, &c.

Q. What country is to the S. of them?

A. A wild country but little known, called Patagonia.

Q. What is the population of these States?

A. Thirteen millions.

CHAPTER L.
Brazil.

Q. What country formerly belonged to the Portuguese?

A. The Brazils, a very large country, being the eastern part of South America, to which the Portuguese Court removed when the mother country was invaded by the French.

Q. What event has happened since?

A. The Brazilians in 1822 declared themselves independent, and proclaimed the eldest son of their former king, emperor of Brazil.

Q. What are the climate and soil of the Brazils?

A. The climate is in general temperate and healthful, and the soil very fertile so far as the country has been discovered, for the inland parts are but little known.

Q. What is the character of the Brazilians?

A. Some of the natives are wild, but others are kind, tractable, and ingenious?

Q. What is the capital of Brazil?

A. Rio Janeiro.

Q. How far is Rio Janeiro from London?

A. 5,723 miles.

Q. What is the population of the empire?

A. Five millions.

CHAPTER LI
Of the West Indies.

Q. What is meant by the West Indies?

A. A certain number of islands on the east side of America, situated near the Gulf of Mexico.

Q. Which are the principal of the West India Islands?

A. The principal are Cuba, Hispaniola, or St. Domingo*, Jamaica, Porto Rico, Barbadoes, Guadaloupe, Martinico, Tobago, St Christopher's, Antigua, Barbuda, Anguilla, Dominica, St. Vincent, Grenada, St. Lucia, with several others of less note.

Q. Why are these islands called the *West Indies*?

A. They are so called from the following circumstance. Christopher Columbus, the discoverer of America, proposed to sail to the East Indies by a western course; but instead of reaching Asia, he found America; still he was persuaded that those islands were the western part of India; and, before a sufficient number of voyages could be accomplished to convince him of his error, he died. From this mistaken notion the islands acquired the name of West Indies; besides to them from England we must sail to the westward; and, as we must sail to the eastward to go to *India*, the terms *East* and *West Indies* are not inapplicable.

Q. What is the population of the West India Islands?

A. Two millions six hundred thousand, of whom only four hundred and sixty thousand are whites.

Q. Are the black population of the British West India Islands, slaves?

A. No; the British Parliament gave them their liberty in 1834, and paid their former masters £20,000,000 sterling as a compensation.

Q. Are there any other islands in the world that we have not mentioned?

A. There are many other islands besides those already described; the principal are New Holland (which is larger than all Europe), Sardinia, Sicily, Corsica, Ivica, Minorca, Majorca, Candia, Cyprus, Rhodes, New Norfolk, New Guinea, Van Dieman's

* Now an Independent Republic of Blacks, called Hayti.

Land, New Britain, New Zealand, &c.; and those
of the Pacific, viz.; the Society and Friendly Isles,
with Otaheite, &c.

CHAPTER LII.

Of the Earth's Roundness.

Q. It is mentioned in Chapter I. that the earth is
nearly round, like an orange ; can you give me any
reason why it is of such a shape ?

A. Yes, because no other figure than that of a
sphere could produce the pleasing and regular succession of day and night, and the variety of the
seasons.

Q. What is meant by a *sphere* ?

A. The earth is sometimes called a *sphere*, sometimes a *spherical body*, and sometimes a *globular body* ;
but, whether we use the terms *globular, spheroid,
sphere*, or *spherical*, it is always to be understood as a
globe, that is, a circular body, which these terms
signify.

Q. How is the earth represented by geographers ?

A. By an artificial globe, on which the various
parts of the surface of the earth are delineated ;
therefore it is to conceive that, if a map of the world
be accurately drawn on-any round body or globe,
the surface of that globe will represent the surface
of the earth.

CHAPTER LIII.

Proofs of the Earth being round.

Q. What proofs have we that the earth is *spherical* ?

A. From viewing distant objects ; for instance, a
ship, going out to sea, gradually disappears to a spec-
tator upon land, and the last part of the ship visible
is the top of the mast ; on the contrary, if a ship be
advancing towards him, the top of the mast comes
first into sight, then the shrouds, and afterwards the
hull, which could not be the case if the earth were
not spherical.   On a plain surface the largest objects
are seen at the greatest distance.

62   CATECHISM OF

Q. What other proof have we?

A. From an eclipse of the moon; for the shadow of the earth falling upon the moon is always circular; which could not be unless the earth was globular; because a body must be a globe, which in all situations casts a circular shadow.

Q. These are proofs, certainly, that the earth must be a globe; but is it not a great mystery how the earth is situated relative to the heavenly bodies, what supports it; and, as it is peopled on all sides, why do not its inhabitants fall from their impending situations?

A. The earth is situated in open space, resting upon nothing, but supported by gravitation, which draws all terrestrial bodies towards its centre; and may be compared to a great magnet rolled in filings of steel; consequently there is no danger of falling from any one side of the earth more than from any other.

Q. What is meant by gravitation?

A. Gravitation is a law in nature, which all bodies have, tending to a central point, called the centre of gravity.

CHAPTER LIV.

Of the Difference and Distance of a Mile in different Countries.

Q. Is the measurement of a mile the same in every country?

A. No, very different, for the English mile is much less than the Indian; and the Indian than the Spanish; the Spanish than the German?

Q. What is an English statute mile?

A. An English statute mile consists of 5,280 feet, or 1,760 yards or 8 furlongs.

Q. What nations agree nearly in this measure?

A. The Turkish, Italian, and old Romish lesser mile, are nearly one English.

Q. What is a Russian mile?

A. Very little more than three quarters of an English mile.

Q. What is an Indian mile?
A. Three English miles.
Q. What is a Spanish, Polish, and Dutch mile?
A. About three miles and a half English.
Q. What is an Arabian mile?
A. A mile and a quarter English.
Q. What is a German mile?
A. Little more than four English miles.
Q. What is a Swedish, Danish, and Hungarian mile?
A. From five to six English miles.

CHAPTER LV.

Of the Equator, &c.

Q. What is the Equator,* and why is it so called?
A. The Equator is a great circle, which divides the earth into two equal parts at an equal distance from each pole; and is so called, because, when the sun is in this circle, the days and nights are equal to all the inhabitants of the earth.
Q. What are the poles?
A. The poles are two points of the earth opposite to each other, the one called North and the other South. These points are only imaginary, as also are the circles.
Q. Why are these points distinguished by the names of North and South poles?
A. Because they are opposite to the North and South parts of the heavens.
Q. Of what use is the Equator?
A. All parts of the earth, with respect to their situations, are either to the North or South side of the Equator; and the distance of places, called their latitude, is counted from it towards the North or South pole.
Q. How far are the poles from the Equator?
A. Ninety degrees, which is the fourth part of a

* The Equator is sometimes called the *Equinoctial Line*, and sometimes only the *Line*.

circle, called a quadrant; half a circle is 180 degrees, and a whole circle, whether great or small, is 360 degrees.

Q. What are the tropics?

A. Two lesser circles called Cancer, and Capricorn. The tropic of Cancer is 23½ degrees north of the Equator, and the tropic of Capricorn 23½ south of the Equator.

Q What are the polar circles?

A. Two lesser circles, called Arctic and Antarctic; the arctic or north polar circle, is 66½ degrees from the Equator. and the antarctic or south polar circle, is the same distance south of the Equator. The polar circles are also 28½ degrees from the poles.

CHAPTER LVI.

Of the Circles of the Globe.

Q. What names are given to those lines or circles that surround the globe?

A. The principal circles which surround the globe are the equator, the ecliptic, the tropics, the polar circles, and the meridians.

Q. What is the ecliptic?

A. The ecliptic is a great circle described by the sun in the space of a year. It surrounds the globe between the two tropics, and crosses the equator at two opposite points, called the equinoctial points.

Q. What are the meridians?

A. All lines drawn from one pole to the other are called meridians. When the sun comes to the meridian of a place, it is then noon or mid-day at that place, for every place has a meridian. The meridian or line, drawn from north to south over Greenwich, is called the first meridian, and the longitude of places is reckoned from it either east or west.

Q. What is meant by the horizon?

A. There are two horizons; one called the sensible or visible, and the other the rational horizon.

The sensible horizon is the boundary of our sight, being that circle where the sky and earth seem to meet. The rational horizon cuts the earth into two equal parts, and is parallel to the sensible horizon.*

Q. What are the poles of the horizon called?

A. The Zenith and Nadir; the Zenith point is that part directly over head; and the Nadir is that point, opposite to it, under our feet.

Q. What are the four cardinal points of the horizon called?

A. North, east, south, and west.

Q. How is a circle divided?

A. All circles are divided into 360 parts called degrees. The half, or semicircle, is 180 degrees; the quarter, or quadrant, is 90 degrees.

Q. How is a degree divided?

A. The degree is divided into 60 minutes, and every minute into 60 seconds.

Q. How many kinds of circles are there?

A. Two, the *great*, and the *less circles*.

Q. What is a *great circle?*

A. A *great circle* is that which divides the earth into two equal parts.

Q. What is a *less circle?*

A. A *less circle* divides the earth into two *unequal* parts.

Q. What is the admeasurement of a degree in a great circle?

A. A degree in a *great circle* contains 60 geographical or 69¼ English miles, but the admeasurement of a *less circle* varies accordingly as it approaches towards the poles.

Q. What is a mile called in geography?

A. A minute.

Q. How many degrees make an hour of time?

A. Fifteen.

* The broad wooden frame which supports an artificial globe represents the rational horizon.

CHAPTER LVII.

Of Latitude and Longitude.

Q. How are places upon the earth distinguished ?

A. By their situations, which are obtained by their latitude and longitude.

Q. What is meant by the latitude of a place ?

A. Its distance north and south from the equator. If it be on the north side of the equator, it is said to be on the north latitude ; if on the south side, in south latitude. All places on the earth are either in north or south latitude, except at the equator, where there is no latitude, because latitude begins there.

Q. What is the greatest latitude a place can have ?

A. Ninety degrees, and there are only two places that have so much, which are the poles.

Q. What are the parallels of latitude ?

A. Parallels of latitude are lines drawn at an equal distance from the equator.

Q. How is the latitude of a place found ?

A. Upon a globe it is found by turning the globe round till the place comes under the brass meridian in which the globe is suspended, which shows the distance from the equator. But upon a map the latitude is found at the side ; if it increases upwards, it is north ; if downwards, it is south.

Q. What is meant by longitude ?

A. Longitude is the distance of a place from the first meridian to the east or west. If it is on the east side, it is east longitude ?

Q. What is the greatest longitude a place can have ?

A. One hundred and eighty degrees, which is one half of the circumference of the globe. All places are either in eastern or western longitude, except under the first meridian, which has no longitude, it being there where longitude begins.

Q. How is longitude found ?

A. Upon a globe it is found at the equator, but upon a map at the top and bottom.*

CHAPTER LVIII.

Of the Zones.

Q. What are Zones ?

A. Certain spaces that encompass the earth like a belt.

Q. How many zones are there ?

A. There are five zones, namely, one torrid, two temperate, and two frigid or frozen zones.

Q. Where is the torrid zone, and why is it so called?

A. The torrid zone includes all that part of the earth which is situated between the tropics, and is denominated torrid, or burning, because of the great and continued heat of the sun, under whose course it lies.

Q. What are the temperate zones, and why are they so called ?

•A. The temperate zones include all those parts of the earth which are situated between the tropics and polar circles; within the two extremes of heat and cold, which renders the air more temperate, on which account these parts are much more improved.

Q. Where are the frigid zones situated, and why are they so denominated ?

A. They are situated between the polar circles extending round each pole, and are called frigid or frozen, from the rays of the sun falling so very obliquely on those parts, which renders it excessively cold.

---

* The degrees of longitude are not equal, like those of latitude, but diminish in proportion as the meridians incline towards the poles. Hence, in 60 degrees of latitude, a degree of longitude is but half the quantity of a degree upon the equator, and so in proportion for the rest.

CHAPTER LIX.

QUESTIONS ON EUROPE.

1. What is the name of the chief city of Sweden ?
2. What is the population of Sweden ? .
3. What is the name of the metropolis of Russia ?
4. What is the name of European Russia ?
5. What is the name of the chief city of Denmark ?
6. What is the population of Denmark ?
7. What is the name of the metropolis of Prussia ?
8. What is the population of Prussia ?
9. What is the name of the metropolis of Holland, &c. ?
10. What is the name of the metropolis of Belgium ?
11. What is the population of the German States ?
12. What is the name of the metropolis of Austria ?
13. What is the population of Austria ?
14. What is the name of the metropolis of European Turkey ?
15. What is the population of European Turkey ?
16, What is the name of the metropolis of France ?
17. What is the population of France ?
18. What is the name of the metropolis of Switzerland ?
19 What is the population of Switzerland ?
20. What is the name of the metropolis of Italy ?
21. What is the population of Italy ?
22. What is the name of the metropolis of Portugal ?
23. What is the population of Portugal ?
24. What is the name of the metropolis of Spain ?
25. What is the population of Spain ?
26. What is the name of the metropolis of England ?
27. What is the population of the British Isles ?

QUESTIONS ON ASIA.

1. What is the name of the metropolis of Asiatic Turkey ?
2. What is the population of Asiatic Turkey ?
3. What is the name of the metropolis of Asiatic Russia ?
4. What is the population of Asiatic Russia ?
5. What is the name of the metropolis of China ?
6. What is the population of China ?
7, What is the name of the metropolis of Japan ?
8. What is the population of Japan ?

9. What is the name of the metropolis of the Birman Empire ?
10. What is the population of the Birman Empire ?
11. What is the name of the metropolis of Siam ?
12. What is the population of Siam ?
13. What is the name of the metropolis of Hindostan ?
14. What is the population of Hindostan ?
15. What is the metropolis of Persia ?
16. What is the population of Persia ?
17. What is the name of the metropolis of Tartary ?
18. What is the population of Tartary ?
19. What is the name of the metropolis of Arabia ?
20. What is the population of Arabia ?

QESTIONS ON AMERICA.

1. What is the name of the metropolis of the United States ?
2. What are their capital and chief ports ?
3. What is the population of the United States ?
4. What were the Spanish dominions in North America ?
5. What changes have taken place ?
6. What new states are there in South America ?
7. What is the population of British America ?
8. What are the chief cities and towns in Canada ?
9. How are the British American Provinces governed ?
10. What is the name of the metropolis of Brazil ?
11. What is the population ?

QUESTIONS ON AFRICA.

1. What is the name of the metropolis of Abyssinia ?
2. What is the population ?
3. What is the name of the metropolis of Egypt ?
4. What is the population ?
5. What is the name of the metropolis of Morocco ?
6. What is the population ?
7. What is the name of the metropolis of Algiers ?
8. What is the population ?
9. What is the name of the metropolis of Tunis ?
10. What is the population ?
11. What is the name of the metropolis of Tripoli ?
12. What is the population ?

THE END

BRITISH STATIONERY;
PLAIN AND FANCY.

—

WRITING PAPERS.

The Subscriber has always on hand a great variety of the following ;—

Whatman's and Wilmot's Superfine, Imperial, Super Royal, Royal, Medium and Demy Writing Papers ; also, Blue and Yellow Wove Folio, large and small Posts, Folio Medium Bank Post, &c.

POST AND NOTE PAPER—Fine and Superfine, Large and Small, Thick Laid, Yellow and Blue Wove Post, of the best makes, Cream Laid Post, Fine and Superfine, Large and Small, Thin Laid Yellow and Blue Wove Posts ; Superfine Small Thick Yellow Wove, Glazed, and Gilt Post ; Superfine Laid Yellow and Blue Wove Note Paper, Gilt and Plain ; Extra Large Thin Post—also, the same of Parisian and Austrian makes for Foregin Correspondence in great variety ; Extra Satin Post Paper, gilt and plain ; Embossed and Plain Letter Note Paper, coloured ; Black Edged and Black Bordered Post and and Note Paper, various breadths.—Post, with Engraved Views of Montreal, Quebec, Kingston, Niagara, &c. &c.

Drawing Papers and Drawing Materials ; Cards and Card Cases ; Steel Pens and Quills ; Wax and Wafers ; Inks and Ink Stands.

BLANK BOOKS.

Consisting of Ledgers, Journals, Day Books, of Superfine and fine paper ; Bill Books, Bank check Books, Ruled Memorandum Books, Pass Books, &c. &c. &c.

PRINTING.

The Subscriber having lately made an extensive addition to his stock of Type, including a great variety of the latest introduced, is prepared to execute all orders for Printing with neatness and despatch, and on terms as moderate as those of any other in the Trade.

BOOK-BINDING.

The Subscriber binds in every variety of Style, Ledgers, Journals, Cash, Day Books, &c. &c.

CHARLES G. DAGG.

PRIZE S[...]

The Subscriber obtained [...]mas [...]
bitious, held at Montreal and Ha[...]
collection of School Books printed and bound [...] Canada."

NATIONAL SERIES.—

General Lesson, to be hung up in Schools.
First Book of Lessons.
Second Book of Lessons.
Sequel to the Second Book.
Third Book of Lessons.
Fourth Book of Lessons.
Fifth Book of Lessons.
First Book of Arithmetic and Key.

English Grammar and Key.
Book Keeping and Key.
Treatise on Mensuration.
Appendix to Mensuration, for the use o Teachers.
Elements of Geometry.
Introduction to Geography a History, with Maps, Pla &c., new edition, much .. roved.

Large coloured Maps for School Rooms.

CURRICULUM LATINUM.

Cornelius Nepos.
Virgilii Georgica.
Cicero de Amicitia.
Cicero de Senectute.
Ovidii Fasti.

Cæsar de Bello Gallico.
Q. Curtius.
Taciti Agricola.
Horatii Carmina.

CANADIAN EDITIONS—SCHOOL BOOKS.

The Canadian Primer.
Manson's Primer.
Mavor's Spelling Book.
Carpenter's Spelling Book.
Webster' Spelling Book.
Walker's Dictionary.
Lennie's Grammar.
Murray's English Reader.
Murray's Large English Grammar.

Murray's Small ditto.
Ewing's Canadian Sch-
    Geography.
Walkingame's Arithmetic.
The History of Canada.
Do. in French.
The History o Rome.
Geography of Canada.
Quarter Dollar Atlas.
Scripture Atlas.

CATECHISMS. &c.

The Shorter Catechism.
The Same with Proofs.
The Mother's Catechism containing common things neces .
    .. be known at an early age.
The .... Catechism, being a Sequel to the First.
The Child's Own Prayer Book.
Catechism for the Instruction of Communicants of the Lord
    Supper, by the late Dr. Andrew Thomson.
Lessons on the Truth of Christianity.
Catechism of Universal History.
Catechism of the History of England.
Catechism of Geography.

CHARLES G. DAGG.

www.ingramcontent.com/pod-product-compliance
Lightning Source LLC
Chambersburg PA
CBHW020232090426
42735CB00010B/1664